A Simple Mathematic Model of Sustainability

SEGUACE SUN

978-1-80541-470-4 (Paperback)
978-1-80541-471-1 (eBook)
978-1-80541-472-8 (Hardcover)

Special thanks to

Chuanqi Wang, Yichuan Guo, Jia Guo and Pav Benning

Contents

Preface

All knowledge from God, His Love, is boundless.
Praise the Lord, praise God.

In quest'ambiente viene coordinato l'insieme della cultura umana con il messaggio della salvezza, di modo che la conoscenza del mondo, della vita, dell'uomo, che gli alunni via via acquistano, sia illuminata dal Vangelo. Alleluia!

In this environment, the whole of human culture is coordinated with the message of salvation so that the knowledge students gradually acquire of the world, life, and man is illuminated by the Gospel. Alleluia!

Walking into the valley of darkness is never easy because there is no light. However, the true light, God, is always shining and guiding me. When I was falling into the darkest valley, the Lord held me and saved me.

During this journey, I saw the beginning of trading and manufacturing and the simple systems involved. As a scientist, I noticed a lot of similarities between this process and mechanics. Therefore, I tried to apply the methodology and concept of mechanics and mathematics to study manufacturing and trading. I found that it worked. Although I do not have a background in economics, I decided to publish this idea and my findings to share them with the world. I understand there may be a lot of points to argue in this book, but the fundamentals should be a good start. I hope brighter people will use this base and enhance it in the future.

At the end of this preface, please allow me to say, Praise the Lord! Alleluia!

赞美恩主, **Alleluia!**
虽堕九泉其志不灭
纵受百死本心未改
千锤百炼彰显本性
浴火涅槃方成正果

希天下众生志心勤苦人人如龙，望文明之光薪火相传奋勇不辍。

赞美恩主！

Lodare Dio, Ti Amiamo, Signore. Ti Voglio Bene Ci illuminate, Signore. Gridiamo al Signore che è salvatore potente. Alleluia!

Praise God, We Love You, Lord. I Love You. Enlighten Us, Lord. Let us cry out to the Lord, Who is a powerful saviour. Alleluia!

Abstract

This book uses the idea and methodologies of mechanics to study the economic loop. Space, time and mass are the fundamental elements of classical mechanics. Here, we replace the distance with the value price and define the demand, which is equivalent to the mass. Then, a simple model of an economic loop is analysed, and the potential and future work of the model are discussed.

CHAPTER 1
Introduction

This is a book to discuss sustainability using the point view of mathematics and Newton's method. We will just change the distance to the value price and discuss the simplest model of an economic loop to analyse the starting point of economics. We will examine why we need sustainability and its key elements, why humans differ from other species, and what we wish to achieve through economic loops. Many ideas in this book are very similar to the *Das Kapital* series, by Karl Marx [1] [2].

The book is divided into several parts:

1. The methodology;
2. The Simplified Isolated System;
3. The simplest model of a loop of economics;

4. The potential hidden risk in economic loops;
5. A discussion on how to achieve sustainability.

This book is just the beginning of this field of study, but my intelligence limits me from doing more and taking it further. Therefore, I hope more people will continue this work in the future.

CHAPTER 2

Methodology

In this book, we will use basic differential and integral calculus to study an economic system model.

2.1 Differentiation and Integration

There are two important parts in calculus: differentiation and integration. They are complementary to each other. Differentiation is used to break down the function into parts, and integration unites those parts to form the original function. [1, 2] Mathematically, differentiation is used to find the slope of a continuous curve, and integration is used to find the area under the continuous curve.

The concept of differentiation and integration can be introduced by the two variables, y and x, in the function of $y = f(x)$

Differentiation: $\dfrac{df(x)}{dx} = \lim_{h \to 0} \dfrac{f(x+h) - f(x)}{h}$

Integration: $A(x) = \int f'(x)dx = f(x) + C \frac{\Delta y}{\Delta x}$

where $f'(x) = \dfrac{df(x)}{dx}$ and C are constant.

Graphs to explain the derivative and integral are shown as follows:

Derivative *Integral*

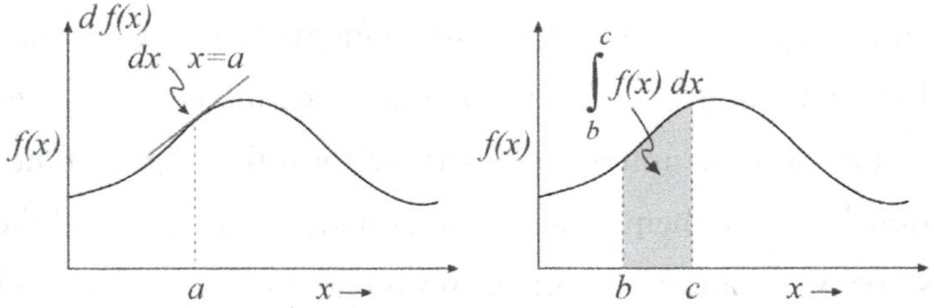

Ref [2]

2.2 The Basics of Newton's Laws

In this book, we have borrowed a lot of ideas from Newton's laws of motion and classical mechanics [3] [5]. The key concepts used in this book are the relationships between displacement (s), speed (v), acceleration (a), time (t) and force (F).

Where m is the mass, their relationships are:

$$F = ma(t) = m\frac{dv(t)}{dt} = m\frac{d^2s(t)}{dt^2}$$

Newton's laws of motion are the basic classical mechanics that describe the relationship between the motion of an object and the force.

1. A body remains at rest, or in motion at a constant speed in a straight line, unless acted upon by a force.
2. When a body is acted upon by a net force, the body's acceleration multiplied by its mass is equal to the net force.
3. If two bodies exert forces on each other, these forces have the same magnitude but opposite directions.

Ref [3] [4]

CHAPTER 3

The Simplified Isolated System

In this model, we borrowed a lot of ideas from classical mechanics. The only difference between classical mechanics and our model is that we have replaced the space with the value price and the mass with the amount of demand. However, the definition of demand is very similar to mass, so the significant change is swapping the space for value.

In our model, the key elements are value, time and amount of demand. Hence, the following assumptions/laws can be made after we reviewed the whole model and system:

3.1 Basic Assumptions/Laws

a. Law 1: A product's value can be estimated by the producer and buyer, but the true value of the product can only be confirmed when the two parties meet.

b. Law 2: The value of a product will remain in a state of constant status unless acted upon by an extra force or demand change.

The following symbols will be used.

Symbol	*Name*	*Unit*
m	demand	usually kg, but depends on product or demand
s	product value	USD/kg
t	time	Second, s
A	space	m^3
n	person/human resource	n (in some situations, this can be a non-integer)
v	Value change speed	USD/(kg·s)
F	Force	C
TD	Total debit	USD
D	Debit	USD/kg
T	Total value	USD
α	ratio of output and input	

The product value (s) [1] is a measure of a product's ability to meet and address customers' needs. In this book, the unit value of s is also shown by the demand or likeability level of a product.

A positive value means people like the product, and a negative value means people hate it.

In this book, we consider all other demands to be rigid, meaning that the product will fully meet the requirements of demand and consumers will buy the entire stock.

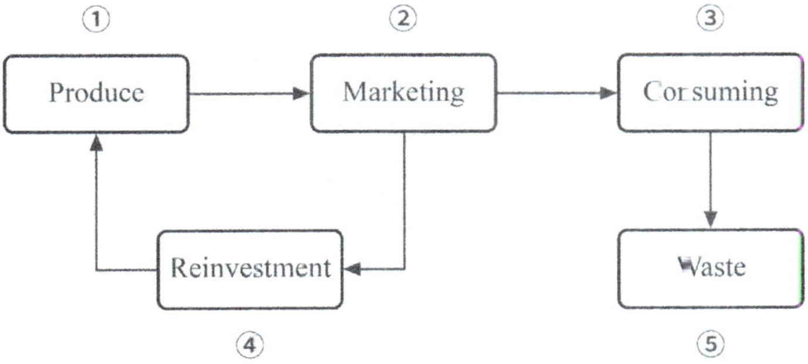

Diagram [1]

The above is a simple version of an economic loop, but they are fundamental and real in economics.

3.2 A Simplified Isolated System

Now, we will have a look at a simplified isolated system, which is a value change process. It can be a simplified process of trading or manufacturing. We assume all the products change at the same time during this process, not one by one. The sketched diagram is as follows:

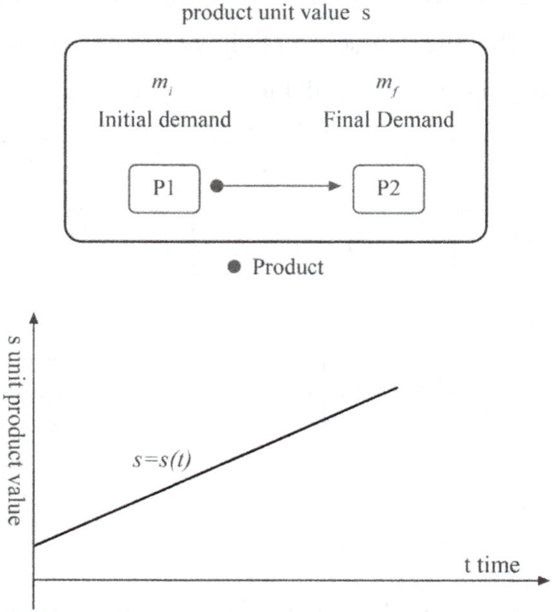

Diagram [2]

m_i is the initial demand for the product, and is the final demand for the product.

m_i can be interpreted as the amount of raw material required related to the amount of the final product (m_f). For example, a gold factory wishes to produce 1kg of gold (m_f), so they may need 100kg of gold ore (m_i).

Therefore, $m_f = \alpha m_i = 1kg$ where $\alpha = 0.01$

If we consider that m does not change, even in a factory,

$$m_f = \alpha m_i = m$$

If we also assume that all of the products change simultaneously during this process, not one by one, there is no waste. All the raw materials will be used completely.

Then, where s is the unit value of the product, the total value is

$$TV(s) = m \bullet s(t)$$

Therefore, the momentum (P) is

$$P = m\frac{ds}{dt} = mv$$

And $v = \frac{ds}{dt}$ is the speed of value change

The value force (F) is

$$F = m\frac{d^2s}{dt^2} = ma$$

And the acceleration (a) of the value is

$$a = \frac{d^2s}{dt^2}$$

Therefore, we can have a similar relationship to classical mechanics, which is:

$$F = ma(t) = m\frac{dv(t)}{dt} = m\frac{d^2s(t)}{dt^2}$$

CHAPTER 4

The Whole Loop

Now, let us go back to the whole loop system. The whole loop can be divided into four main stages: the Manufacture stage, Trading stage, Consumer stage and Reinvestment stage. For the purposes of this book, we consider the whole loop to be stable and running; its number is the o-th cycle, where o is an integer. An example will be shown at the end of the chapter. The diagram can be sketched as follows:

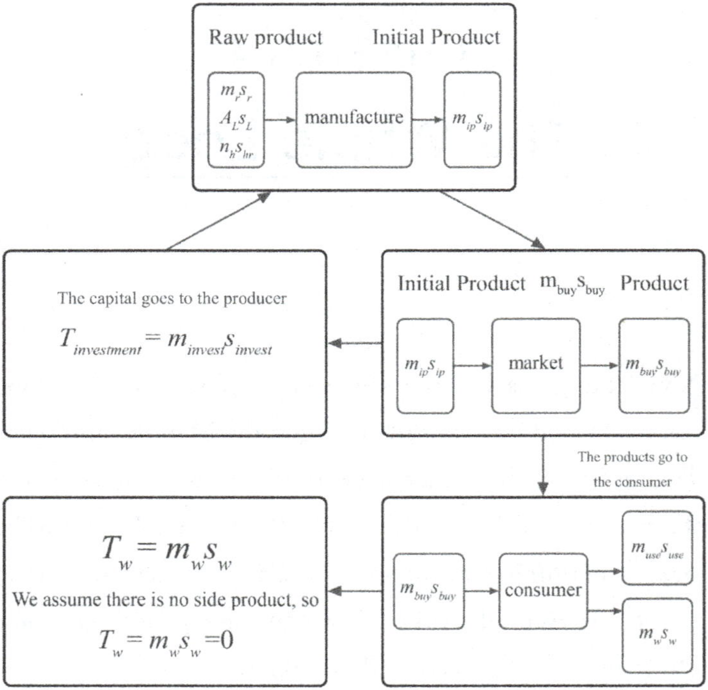

Diagram [3]

s_{ip} is an estimation of the initial product value, based on all costs. s_r is also an estimation from a previous or external loop.

We need to make some assumptions about this system:

1. Assume it is rigid demand and has been sold completely; therefore

$$m_{ip} = \alpha m_r = m_{buy} \tag{1}$$

2. Assume the product will be used completely.

$$m_{buy} = m_{use} \tag{2}$$

A side product or non-product output is the output which does not meet any demand in this loop, including waste and other things. Waste means a negative value compared to demand. In this simplified loop model, we assume there is no side product or non-product output.

We have just looked at a very simple model of a product here; it only needs to be manufactured once and sold once to arrive at the end user. We also did not consider the other extra costs, such as tax, accountant, logistics, etc. But all of these elements can easily be added into the system. The researcher who follows me can take this next step. This kind of product can be gold, rice, etc.

4.1 Stage 1: Manufacture Stage

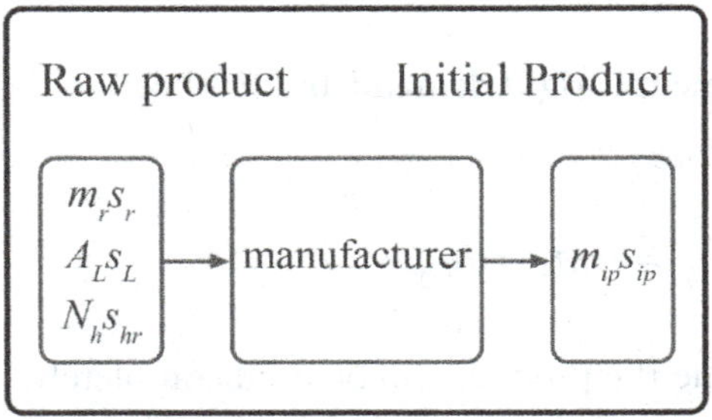

Diagram [4]

The initial raw material value is

$$T_{initial} = m_r s_r \tag{3}$$

If we assume that the demand doesn't change, all the raw material becomes part of the product and there is no waste in the loop, then

$$m_{ip} = m_{buy} = \alpha m_r = m = constant \tag{4}$$

This is when we use α=1 in this loop.

Then, we have

$$T(t) = m \bullet s(t) \tag{5}$$

However, to make a working loop, three key elements need to be invested. These elements are:

a. Consumable values/raw material value, $T_{initial}$, include energy, raw material, water, etc.

$$T_{initial} = m_r s_r$$

b. Non-consumable items: equipment, space, etc. In a loop, this can be measured by the value of space, including space, land, ocean, sky, etc.

Units are measured in cubic metres (m^3) and the unit value s_L states the value per space m^3, such as the machinery required, infrastructure in the space, office rent, etc.

So, the non-consumable value (N_c) in a loop is

$$N_c = A_L s_L \tag{6}$$

If there is any change in the space, then

$$N(t) = A_L s_L(t) \tag{7}$$

s_L is the value change or rental cost, per product, per unit space in a loop. We call it the rental cost because we rent everything from nature. A_L is the total space.

c. Human cost, the labour cost: the total number of people times their labour cost in this loop, so

$$H_c = \sum_n s_{hr,n}$$

Here, $n = 1, 2, 3\ldots$, but n can also be a non-integer, for example, if the industry does not need a full-time employee.

If we only consider each of the labour costs to be the same, then

$$H_c = n_h s_{hr} \tag{8}$$

$$H(t) = n_h s_{hr}(t)$$

Therefore, the total value of the initial product (T_{ip}) is

$$T_{ip} = T_{initial} + N_{ci} + H_c$$

And the total initial capital (D_{Fi}) is

$$D_{Fi} = T_{ip} = T_{initial} + N_{ci} + H_c \quad (9)$$

$$D_{Fi} = T_{ip} = m_r s_r + A_L s_L + \mathrm{n}_h s_{hr}$$

In this book, the raw material costs never change in a loop.

Therefore, at the first stage,

$$T(t) = T_{initial} + N_{ci}(t) + H_c(t)$$

At this stage, the value of the initial product cannot be calculated, because of Law 1 mentioned in section 3.1: the value can only be finalised when the buyer and seller meet.

The seller can have an expected value based on observations from previous cycles.

4.2 Stage 2: Trading Stage

During the marketing or trading process, the product is no longer modified physically. Although logistics, tax, accountancy, retailer fees, waste costs and the other extra costs can be easily added into the manufacturing part, here, we are only considering the simplest model: buyer and seller. Therefore, we can sketch the simplest version of this diagram as follows:

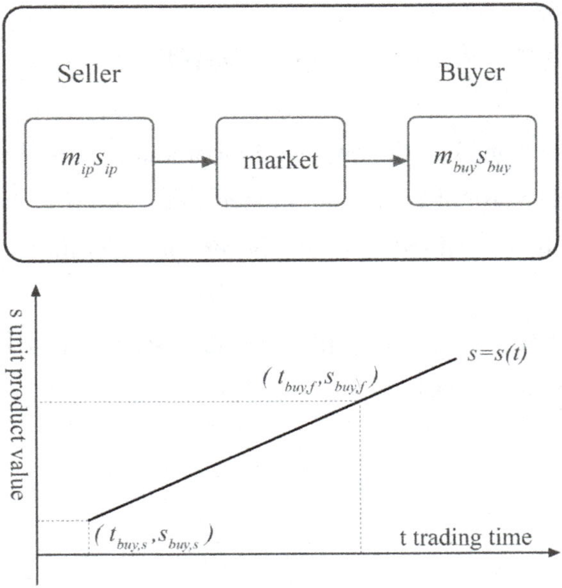

Diagram [5]

In this book, we assume the product meets the user's demand perfectly: the direction of the product perfectly overlaps with the direction of the user. However, the direction of value will not be discussed.

Using equations (1) and (2), we have

$$m_{ip} = m_{buy} = m_{use} = m \tag{10}$$

The trading time is also an important element of the final deal. It allows the suitable buyer time to find the product, strike a bargain, etc.

Therefore, the value of the product (s) in the trading stage is related to the time:

$$s = s(t) \tag{11}$$

The speed of value change and the acceleration is

$$v = \frac{ds(t)}{dt} \tag{12}$$

$$a = \frac{d^2 s(t)}{dt^2} \tag{13}$$

The force of value F is

$$F = m\frac{d^2 s(t)}{dt^2} \tag{14}$$

In this book, we assume all of the products are sold at the same time.

Based on the above, the cost of the product is s_{ip} and the final value for the buyer is s_{buy}.

Therefore, the unit profit (r) is

$$r = \Delta s = s_{buy} - s_{ip} \tag{15}$$

Where t_i and t_f are the time, the relationship between r and t is

$$r = \Delta s = \int_{t_i}^{t_f} v dt = \int_{t_i}^{t_f} \frac{ds(t)}{dt} dt \tag{16}$$

Where m is the constant, the final total profit R is

$$R = m \bullet r(t) = m \bullet \int_{ti}^{tf} \frac{ds(t)}{dt} dt \tag{17}$$

4.3 Stage 3: Consumer stage

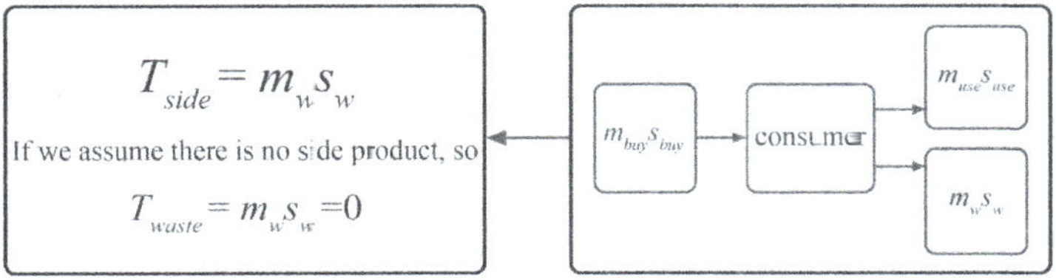

Diagram [6]

The consumer stage is a simple process. We assume the consumer used all of the products from the market so the side product is $T_{side} = 0$.

Therefore,

$$m_{buy} s_{buy} = T_{use} + T_{side} \tag{18}$$

$$m_{buy} s_{buy} = T_{use} + m_w s_w$$

If $m_{w,f} = 0$, there is no side product,

$$T_{use} = m_{use} s_{use} = m_{buy} s_{buy} \tag{19}$$

All of the products are consumed at the same time, and using equations (1), (2) and (10), we have

$$T(t) = m \bullet s(t) \tag{20}$$

The consumer's force (F) is

$$F_{use} = \frac{d^2T(t)}{dt^2} = \mathrm{m}\frac{d^2s}{dt^2} = ma \tag{21}$$

The demand of m_{use} is the origin of all of the other demand.

4.4 Stage 4: Reinvestment

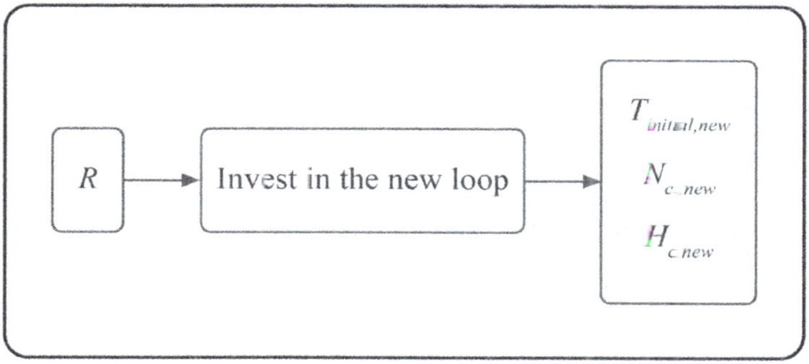

Diagram [7]

To continue this loop, reinvestment is a vital process. Using equation (9), we have the total initial capital (D_{Fi})

$$D_{Fi} = T_{ip} = T_{initial} + N_{ci} + H_c \tag{22}$$

Here, we consider that N_{ci} and H_c are constants; hence, after the trading stage, the total capital (D_F) is

$$D_F = T_{buy} = T_{initial} + N_{ci} + H_c + R \tag{23}$$

After the trading stage, the manufacturer gets the profit. As we mentioned, we are only considering that there is no side product and waste is zero value and not a negative value.

Therefore, for the next cycle, the o+1 cycle, we have the new initial capital, which is

$$D_{Fi,o+1} = T_{initial,o+1} + N_{ci,o+1} + H_{c,o+1} \tag{24}$$

The investment (I) is

$$I = D_{Fi,o+1} - D_{Fi}$$

We are not considering any other external elements here, so all of the profit (R) will be reinvested into the next cycle. Therefore,

$$R = I = D_{Fi,o+1} - D_{Fi} \tag{25}$$

The investment (I) can be used to change $T_{initial}$, N_{ci}, and H_c

Therefore, we can have three options:

If I=0, the loop will remain stable, not increase or decrease.
If I<0, the loop will shrink until the loop disappears.
If I>0, the loop will increase and grow.

4.5 The Whole Loop

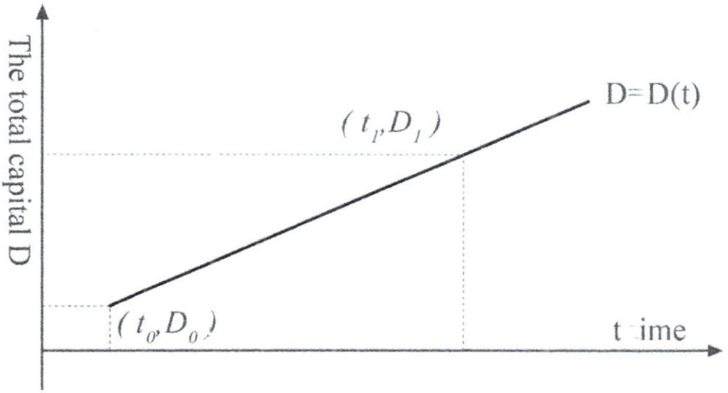

Diagram [8]

Now, we can calculate the total input and gain from a cycle in this loop. Using equations (22) and (23), we have

$$D(t) = T_{initial} + N_{ci}(t) + H_c(t) + R(t) \qquad (26)$$

Where $R(t) = 0$, except at the stage of trading, $T_{initial}$ is a constant in a cycle, and $N_{ci}(t)$ and $H_c(t)$ can be considered as constants in a cycle.

Therefore, the change in the amount of the total capital (p_d) is

$$p_d = \frac{dD(t)}{dt} = \frac{d(T_{initial} + N_{ci} + H_c)}{dt} + \frac{dR(t)}{dt} \quad (27)$$

Here, p_d is not the speed.

Using equations (16) and (17)

$$p_d = \frac{dD(t)}{dt} = \frac{dR(t)}{dt} = m\frac{dr(t)}{dt} = m\frac{ds(t)}{dt} = mv \quad (28)$$

Hence, the force of total capital (F_d) is

$$\begin{aligned} F_d &= \frac{d^2D(t)}{dt^2} = \frac{d^2R(t)}{dt^2} \\ F_d &= m\frac{d^2s(t)}{dt^2} = m\frac{dv(t)}{dt} \end{aligned} \quad (29)$$

Therefore, we can have the force of the loop (F), which is

$$F = F_d = \frac{d^2D(t)}{dt^2} = m\frac{d^2s(t)}{dt^2} \quad (30)$$

Now, we will analyse an example to better understand the model. We will use the process of origami.

Origami is simply the art of folding paper. A paper crane is a very normal shape. We can take this as an example. One small piece of paper can be used to make a paper crane It does not change any physical or chemical properties, or have any side products, and one raw material can become one crane ($a = 1$). However, after taking into consideration people's work, its value can change a lot.

CRANE TSURU

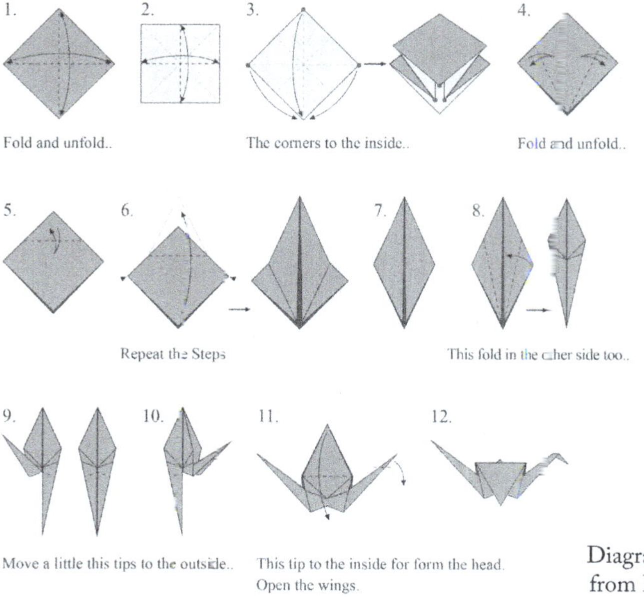

Diagram [9], from Ref [5]

For example, one worker needs to fold m pieces of square paper to m paper cranes. It takes t_1 second(s) to make m paper crane. This worker needs a $20m^3$ office in which to make it. The cost of the rent and furniture in this office is s_L in USD. The salary of the worker is s_{hr}. Therefore, where A_L is $20m^3$ and $n_h = 1$, the total capital in the first stage is $D_{Fi} = T_{ip} = m\ s_r + A_L s_L + n_h s_{hr}$

So

$$D_{Fi} = T_{ip} = m\ s_r + 20 s_L + s_{hr}$$

At the trading stage, the worker sold the m pieces of paper cranes using t_2 seconds with the price of s_{buy}. Then, the profit (R) is

$$r = s_{buy} - s_r \text{ and } R = m(s_{buy} - s_r)$$

The average speed (v) is $v = \frac{r}{t_2} = \frac{s_{buy} - s_r}{t_2}$

The total capital is

$$D_F = m\, s_r + 20s_L + s_{hr} + m(s_{buy} - s_r) = m\, s_{buy} + 20s_L + s_{hr}$$

At the reinvestment stage, the worker uses the total capital to begin the new cycle.

CHAPTER 5
Discussion

In this book, we have only discussed the simplest model of economic loops: no waste at all, and all of the demands are rigid. A real-life situation may be much more complicated, but we are still able to learn a lot from this model.

5.1 Discussion about the Model

With this model, we can easily get the value force (F), capital change (D) and investment (i) of each cycle.

Using equations (22) to (30), we have

$$i = D_{Fi,o+1} - D_{Fi} = m \bullet \int_{ti}^{tf} \frac{ds(t)}{dt} dt \tag{31}$$

However, all of the demand of m_{use} is the origin of all the other demands. The capital renews based on the user's demand at every cycle.

Therefore, there are two conditions that decide the upper limit of the loop, which are:

1. The maximum of the user's demand. This determines the market size and the biggest capital size.
2. The supplier of raw materials. This decides the maximum quantity of the product.

There is also another restriction, which is the maximum capacity of waste, but we are not considering waste in this book.

The investment (i) is the bridge to balance the supplier's ability and the user's demand.

If i=0, the loop will remain stable, not increase or decrease.
If i<0, the loop will shrink until it disappears.
If i>0, the loop will increase and grow.

Here, we also ignore the phenomena of scale to simplify the model, but this can be considered easily in any further study.

In the real world, many loops form the economic network. The whole economic capital needs to grow. It is because when the economy grows, that means most of the loops are growing. However, once a loop is saturated or shrinks, the effect will be transferred to the next until new loops and networks are built.

Therefore, we have to continue to build more new loops to find more resources to meet more needs. Using this model, we can understand that demand and raw materials are key elements to have a sustainable economy. Hence, the entire economy should keep developing new loops and technology to find more resources and increase demand.

On the other hand, if we consider the waste in the whole system, we can also find that waste is one of the biggest ways of losing capital. So, the waste needs to be managed and reduced. Luckily, the waste from one loop can be the resource of the other loops, so a good economic network should work with each other to reduce the total society waste.

5.2 Further Work

This is only the start of the work that can be done here. Using this model, we can see there are a lot of similarities between mechanics and economics. This model has great potential for development, and a lot of methodologies in mechanics can be used in this field, meaning it can be developed quickly. However, there are a lot of discussions and validations that need to be made in academia.

For example, during the trading stage, what is the best explanation of the market force? Many ideas can be borrowed from mechanics, such as gravity or a Two Spring Oscillator system. If we consider the concept of market force similar to the Two Spring Oscillator system, then we can have the following:

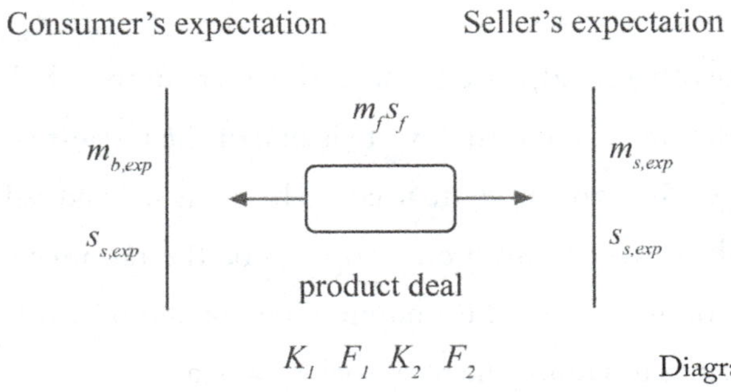

Diagram [10]

$$(m_{b,exp}s_{b,exp} - m_f s_f) \cdot K_1 + (m_f s_f - m_{s,exp}s_{s,exp}) \cdot K_2 = 0 \quad (32)$$

$$F_1 = (m_{b,exp}s_{b,exp} - m_s s_f) \cdot K_1 \quad (33)$$

$$F_2 = (m_f s_f - m_{s,exp}s_{s,exp}) \cdot K_2 \quad (34)$$

Here, F_1 is the buyer's market force, F_2 is the seller's market force, and K is the buyer's and seller's market constant.

If $m_{b,exp}$ = $\mathrm{m}_{s,exp}$ = m_d, then

$$K_1 \cdot m_d (s_{b,exp} - s_f) + K_1 \cdot m_d (s_f - s_{s,exp}) = 0$$

$$K_1 = (s_{b,exp} - s_f) + K_2 (s_f - s_{s,exp}) = 0 \quad (35)$$

K is the constant of market force.

5.3 The Relationship with the Bible

This model was originally inspired by classic physics, the simplest market in society and the Bible. When I was writing this book, I always asked where the first loop came from. And where was the first capital from? Why do we need sustainability?

It is even possible to consider the first loop and first capital being from Eve's apple [6]. And all of us are like the workers in the vineyard [6] (Matthew 20:1-16) because all physical materials are from nature, from God. We did not create them.

However, we used them to satisfy our demands. Therefore, we should try our best to generate the best fruit for the world [6] (John 15:2-6).

CHAPTER 6
Conclusion

In this book, we borrowed the methodologies and ideas from calculus and mechanics to study economics. The fundamentals of classical mechanics are space, time and mass. Here, we replaced the distance with the value price and defined demand, which is equivalent to the mass. Then we analysed a simple model of an economic loop and discussed the potential and future work of the model.

References

[1] K. Marx, Capital: A Critique of Political Economy, Volume 1, Penguin Classics, 1990.

[2] K. Marx, Capital: A Critique of Political Economy, Volume 2, Penguin Classics, 1992.

[3] M. M. Woolfson, Mathematics for Physics, Oxford University Press, 2007.

[4] "https://www.cuemath.com/differentiation-and-integration-formula/," 23 oct 2023. [Online].

[5] H. D. Young, University Physics with Modern Physics, Pearson, 2023.

[6] "http://hyperphysics.phy-astr.gsu.edu/hbase/Math/derint.html#:~:text=The%20derivative%20can%20give%20you,as%20a%20function%20of%20x.," 27 Oct 2023. [Online].

[7] "https://www.nps.gov/mapr/learn/origami-cranes.htm," 29 Oct 2023. [Online].

[8] CEI Conferenza Episcopale Italiana, La Bibbia di Gerusalemme, Edimedia, 2008.

[9] Wikipedia, "https://en.wikipedia.org/wiki/Classical_mechanics," 29 Oct 2023. [Online].

Made in the USA
Monee, IL
04 November 2026

67380104R00030